ACKNOWLEDGEMENTS

I should like to thank my many colleagues in both university and industry for their helpful suggestions in the preparation of this work, and in particular Professor S. S. Gill, Dr J. Parker, and Mr A. Kelly for their encouragement and support.

The case study in Appendix 3 was carried out with reference to equipment manufactured by Chloride Technical Ltd, who kindly gave their permission for the results of the work to be published. The active cooperation of the Company's engineers in that research is gratefully acknowledged.

Graham Thompson

Design Review

THE CRITICAL ANALYSIS OF THE DESIGN OF PRODUCTION FACILITIES

Dr G. Thompson
Department of Mechanical Engineering
UMIST

Process Industries Division
Institution of Mechanical Engineers

Mechanical Engineering Publications
London

© G. Thompson 1985

All rights reserved. No part of this publication may be reproduced, stored in a retrieval system, or transmitted in any form or by any means electronic, mechanical, photocopying, or otherwise, without the prior permission of the publisher.

ISBN 0 85298 578 9

The Publishers are not responsible for any statement made in this publication. Data, discussion, and conclusions developed by authors are for information only and are not intended for use without independent substantiating investigation on the part of potential users.

Printed by The Lavenham Press Ltd, Lavenham, Suffolk.

Contents

	Introduction	1
	The requirement for a design review	1
	Life-cycle costs	2
	Readership	3
Chapter One	Design Review Experience	4
1	Economic justification	4
2	Quality assurance	5
3	The design review method	6
4	Summary	9
Chapter Two	Objectives	12
1	Statement of objectives	12
2	Levels of design	13
Chapter Three	System Review	15
1	System review prior to detail design	15
2	System review after detail design	16
Chapter Four	Functional Unit Evaluation	20
1	Introduction	20
2	Utility value	21
3	Machine analysis	22
4	System utility	22
5	General considerations	23
Chapter Five	Component Analysis	25
1	Scope of analysis	25
2	Analysis method	25
3	Application of the analysis methods	27
Chapter Six	The Structured Design Review	28
1	Procedure	28
2	Team composition	29
3	Management and control	31

Appendix 1	Maintainability prediction	34
Appendix 2	System utility	37
Appendix 3	Functional unit evaluation: case study	43
Appendix 4	Component analysis examples	58
References		63

Introduction

THE REQUIREMENT FOR A DESIGN REVIEW

Good design is a pre-requisite for the success of any project in which new production facilities are to be built. Design teams have an extremely difficult job to do. Firstly, an overall scheme has to be devised that will actually perform the required operation, then individual items have to be selected, adapted, or designed to carry out specific tasks. In addition to ensuring that equipment will carry out required functions and not create an unacceptable safety hazard, costs have also to be considered. The profitability of a production facility depends upon the capital cost, operating costs, and, finally, on the disposal cost when its useful life is finished. Should difficulties arise during the operation of a plant, it is obviously not feasible to undertake a comprehensive re-design. Substantial improvements may be made during commissioning and during the life of a plant, but the initial scheme will limit the type of changes possible. In fact, once a decision to build has been taken, an upper bound is effectively placed on the future profitability of a plant through the determination of capital cost, production capability, and basic elements that affect operating costs.

There is, therefore, a clear incentive to ensure that the proposed design of any production facility is as good as can be reasonably expected. Such a review of design proposals is the subject of this book.

Different companies have different design procedures, as their businesses dictate. Indeed, what is suitable for one may be entirely inappropriate for another. Priorities in design also vary: a company which, say, handles hazardous materials should incorporate reliability calculations with the synthesis stages of design; another may never think of quantitatively considering reliability at all. Approaches to the undertaking of a design review will naturally vary from company to company, and it is not possible to lay down a rigid design review procedure for universal application. However, it would be impracticable to carry out a worthwhile analysis of substantive design proposals without any kind of plan. The following chapters develop a systematic design review procedure which may be adopted in whole or in part to suit different needs. Certain exercises proposed as part of a design review may even be seen by some as an integral part of other design activities. The design review is considered at all stages of design, from the plant specification through to the analysis of components. The flexibility of approach is evident also in specific analysis methods. For example, a method is described for machine evaluation in which it is possible to make trade-offs between variables in order to optimize and attain a maximum value to the system of which the machine is a part. A case study illustrates the method.

A brief mention of terminology is appropriate at this point. The term 'design review' has been adopted, but in some publications 'design audit' is used. Both terms are in common usage to describe the same kind of design analysis exercise. The term 'review' is preferred to avoid the more rigid connotation of the term 'audit'. One may reasonably assume that an audit exercise is a strict checking exercise against a specification, whereas the kind of analysis proposed is a synergistic one which complements the original design activities. However, in Chapter One where the literature is reviewed, the terminology used by the respective authors is retained.

LIFE-CYCLE COSTS

It is worth considering briefly two investigations which emphasize the importance of life-cycle costs and add weight to the case for a design review. In 1970 it was reported that the direct cost of maintenance engineering in the manufacturing industries was approximately £1100M **(1)**. When building and plant maintenance in nationalised industries were included, this figure rose to over £3000M. A more recent survey by the Centre for Interfirm Comparison shows that maintenance costs are still a significant factor **(2)**. For British manufacturing industry (excluding the nationalised industries), the estimated annual maintenance bill is put at £3000M–£4000M. The level of annual maintenance costs averaged 14.9 per cent of the value of the maintainable assets (replacement value). Thus, the total maintenance cost alone over the life of a plant may well exceed the initial capital cost.

Whilst maintenance costs may, therefore, be very significant, such costs will not be predictable in a precise form at the design stage. This does not mean that only cursory attention can be paid to this aspect of design. The variables which directly influence maintenance costs may be dealt with quantitatively. Thus, the maintainability and reliability of equipment is always of importance and should always form part of a design review.

Introduction

READERSHIP

Published work on the design review is confined to various papers presented at conferences and some published in technical journals. Several books touch upon the subject briefly, but only in so far as their principal subject is involved with the design review, e.g., reliability, maintenance management. This text correlates published literature (fully referenced) with original work by the author, and has the design review as its central theme. The intended readership is the professional engineer and the advanced student of engineering design.

The background subject matter is developed to give the concepts a sound academic base, whilst the practicalities concerning the use of the design review are fully considered. The text may be used by engineers wishing to initiate a design review or to develop existing practice into a more thorough procedure.

A continuous presentation is used from a review of experience, through a formal statement of objectives and analysis methods to a summary of the design review procedures. Academic and practical aspects of the work are in general integrated. However, in order to sustain the continuity of the main text without the distraction of substantive theoretical and case history digressions, extensive use is made of appendices.

Appendix 1 gives the theoretical basis for maintainability prediction and outlines basic reliability techniques, and the academic foundation and use of the system utility concept is discussed in Appendix 2. Appendix 3 describes fully a case study at functional unit level, and examples of component analysis are given in Appendix 4.

Thus, the academic and practical aspects of the design review are both considered. The main text should be treated as 'core material'; the reader not wishing to study background theory could then omit Appendices 1 and 2, whilst the student of engineering design should study both the theory and practice given in the appendices.

Chapter One
Design review experience

1 ECONOMIC JUSTIFICATION

Some organizations already adopt a form of design assessment, but it is difficult to gauge the extent to which successful review procedures are being implemented. However, a number of studies have been reported, and a critical look at this experience provides a useful basis upon which to develop a more definitive procedure.

The design review is basically an exercise that formally checks that a given design is suitable for its intended purpose, and meets any specifications that may have been laid down. The concept is not new, of course; it must have existed since man has expressed ideas on paper rather than transforming them directly into hardware. Design evaluation can take the form of a systematic strategy or a general view of the available information, looking for problems as the data is scanned. If a systematic strategy is adopted, it is likely that more costs will be incurred at the design analysis stage than if the informal approach is used, due to the extra time involved. Usually there is pressure to commence the construction phase of a project and, consequently, the informal design appraisal may be curtailed before many potentially serious problems have been identified. Once set in motion a formal procedure will stand a better chance of running its full course before being stopped.

There is a need to justify the design review on economic grounds. Smith (**3**) gives a breakdown of costs over a six month period for an organization concerned with light machining, assembly, wiring, and functional testing of electrical equipment. Sales over the study period were £2M. The analysis of new designs prior to the release of drawings plus audits of systems, products, and processes amounted to £2900. Of failure costs, design changes accounted for £18 000, commissioning failures £5000 and fault finding £26 000. The last item is not only a function of the fault finding techniques used, but also of the number of faults occurring. Smith goes on to deduce that a 10 per cent reduction in failure costs would release approximately £500M into the UK economy. The potential cost savings clearly justify a significant increase in the relatively small costs incurred at the design evaluation stage. In the process industries, Jeffries *et al*. (**4**) claim that a design audit need not necessarily lead to an increase in the initial cost of plant. They argue that a reappraisal of a design may actually reduce initial costs. Also, since capital costs are a relatively small proportion of life-cycle costs (about 15 per cent for a steel plant with a life of 15 years), substantial benefits can be achieved by directing attention to facets of the design affecting future maintenance operations. Jeffries and Westgarth (**5**) cite two cases where a design audit has been carried out: a component audit of the main ram cross head of an extrusion press, and a system audit of a new materials handling complex. An annual cost saving of £380 000 against a capital investment of only £56 000 was estimated in the latter case. The economic case appears therefore to be justified, with particular emphasis on maintenance and life-cycle costs being warranted in process industry applications.

2 QUALITY ASSURANCE

The design review is a significant part of quality assurance systems. BS5750 *Quality Systems*, Part 1 (1979), is concerned with specification for design, manufacture, and installation. It states that the supplier shall maintain the control of certain design functions, one of which is the establishment of design review procedures. The object is to ensure that the design and development programme achieves its objectives by identifying problem areas before manufacture. The standard does not, however, give details of how the review should take place. It is important to note that any supplier claiming to conform to the British Standard must establish a design review procedure. This applies equally to equipment supplied to a large scale process plant as it does to consumer products. Baker (**6**) considers that a design review meeting is an important aspect of quality assurance. He gives a list of major review considerations, which include maintainability, reliability, and maintenance policy. Again, no details of how the review would be carried out are given, except that the design project manager should chair the review meeting.

Napier (**7**) divides design into three main elements: creativity, management support, and review. Quality assurance is emphasized as being part of the design review. Review is defined as the formal, documented, and systematic study of a design by the user and by specialists not directly associated with it. He argues that it should not be 'design by committee'; instead, it is a formal opportunity to monitor the progress being made to satisfy the design requirements. The role of the quality assurance engineer, acting as a bridge between design and production, is said to be important. A proviso is made that the person concerned must become involved in matters which are currently not considered part of quality assurance. This comment is interesting since, if the idea was applied to the construction of production facilities, the quality assurance engineer would usurp the role of project engineer.

3 THE DESIGN REVIEW METHOD

Lawler **(8)** discusses a two stage design review as the most effective way of monitoring a design. The first review deals with the design concept before the main design/development work commences. The final review is held prior to the manufacture stage. It is suggested that the reviews should be formally documented. A checklist is given which includes reliability, performance, maintenance, manufacture, and safety. The review panel, chaired by an engineer, would consist of the following: designers of the system under review, other senior designers, quality assurance engineers, a production engineer, and a material control representative. The chairman should not be the immediate superior of the persons whose work is under review. This mix of expertise is clearly not intended for the analysis of a large production facility, but some points are of direct relevance. The relationships between the chairman, designers of the system under review, and the addition of other design staff is interesting. Apart from a checklist of some twelve points for consideration, and a detailed flow chart for management control, no methods are suggested as to how the team judges whether a design is satisfactory.

A brief case history of a design review activity used by Lansing Bagnell has been published **(9)**. No details of the product are given, but the general form of the review process is described. A three stage review has been adopted. The preliminary review is carried out after a significant proportion of the design has been completed and when cost estimates have been prepared. This differs from the practice described by Lawler where concepts are first discussed. The second stage review is carried out during the hardware development stage. The final review is to confirm the effectiveness of actions resulting from the second review and to finally resolve any outstanding problems. The review panel composition is not discussed in detail, but it is suggested that design engineers and representatives nominated by each department head should comprise the team. In a design review of a production plant, the second and third reviews of the Lansing Bagnall practice seem inappropriate since large amounts of money will have already been committed. The two stage approach given by Lawler is more useful for a production plant review, since the first stage aims to ensure that the concepts are properly established and the second checks that the detail design will put the concepts into practice.

Jeffries *et al*. define the design audit as: 'The technical vetting of plant to ensure that it is reliable, safe, easily maintained, and operating with minimum downtime.' This is a good definition that encompasses the general aims of designers. However, the inclusion of a reference to downtime appears unnecessary since reliability and maintainability have been mentioned. An audit in three parts is described: total system audit, functional unit audit, and component audit. The methods by which these audits may be carried out are not given, but the expected results of the exercises are described. The total system audit aims to highlight critical areas of plant which may threaten availability (reliability modelling is suggested to appraise and modify plant layout). The total system audit is similar to the first conceptual stage proposed by Lawler. The functional unit audit is a detailed one in which the design of individual plant items are compared against a checklist which typically includes stress tolerances, maintainability, and reliability. No analysis methods are mentioned. The object of the component audit is to consider such details as wear rate, maintenance skill requirements, and compatibility of parts. Clearly this level of analysis is impracticable for the whole of a large scale plant unless tremendous personnel resources are available. The authors do not suggest how components may be identified for inclusion in this audit. The three levels of audit fit into a three stage programme, given as: familiarization, criticality study (the total system audit is carried out at this stage), and finalization and recommendations (functional unit and component audits). It is noteworthy that the need for specialist expertise in carrying out a design audit is established; examples are: lubrication engineers, material technologists, and engineering economists.

A design review procedure has been adopted by the US Army **(10)**. A project is reviewed at five stages.

(1) The first review is held when the contractor (or in-house design team) receives the specification to ensure that the requirements are properly understood.

(2) The proposed concept is reviewed to make certain that the project does not contain too many high risk features.

(3) Upon the completion of the design, and prior to the release of drawings for development, the design characteristics are analysed in detail to confirm that the initial requirements are met.

(4) and (5) These involve, respectively, prototype and service test reviews, and make use of extensive checklists. Very detailed aspects of the equipment are considered, referring to the provision of suitable adjustment and test points, a maintenance plan, and support requirements. The specific points identified apply mainly to electrical appliances. A typical question is: 'Has protection been provided for cabling around sharp corners?' The yes/no answer is a feature of the checklist, and equipment subjected to this detailed analysis ought to have a good service record.

Although five stages are given in this review process, the same basic levels are used for design analysis as reported in other procedures: the systems level (conceptual), functional unit or product level (design characteristics), and a detailed component check.

The design review method applied to new product development is supported by Levesque **(11)**. Suggesting that new products '. . . are frequently designed out of hand with little attention given to alternative product forms of composition', he proposes that component synthesis using a morphological chart should be carried out as part of the audit. The morphological chart is an established design method used to generate alternative solutions **(12)**. Its incorporation in an analysis exercise can be supported if it is used to demonstrate a lack of imagination in the initial design but not to quantify the quality of a design. However, the audit procedure is said to improve: product performance, cost reduction, product life, and safety, and to achieve unique product features. Lavesque asserts that it is unrealistic and uneconomic to consider all these goals when analysing an existing product, but that all can be considered in new product development. Further discussion of the role of the design review in new product development, from market research to manufacture, is given by Jacobs and Mihalsky **(13)** and Pugh **(14)**. Jacobs and Mihalsky give a detailed design review procedure and a list of participants in the design review. However, the comments made are not directly transferable to the design of production facilities. For example, maintenance is noted only as one of the several responsibilities of a field engineer. Pugh proposes a six stage review which is integrated within the product development programme from market research to manufacture. Much emphasis is placed on the product design specification as the basis for the design review process. Also of particular interest is the distinction made between the activity of a design review and the resources required to implement the actions arising from the review. The need for specialist expertise to complement the design review team was also established by Jeffries *et al*.

4 SUMMARY

A pattern has emerged with respect to the design process. (The evaluation stages after manufacture will be omitted as being inappropriate for the design review of high capital investment facilities.) Firstly, there is a need to establish that the contractor, or in-house design team, fully understands the implications and relative importance of the various parts of any design specification. Once the initial scheme has been established, a system review should take place to identify any fundamental weaknesses. Reference may be made to plant availability or decisions taken concerning the perusance of high risk novel designs. The US Army review procedure in particular emphasizes the latter point. A detailed examination of the design of 'functional units' or 'products' needs to be carried out next. The most detailed evaluation takes place at the component level. Clearly, this needs to be selective for a large scale production facility with certain types of components being singled out for attention. The four levels of a design audit are therefore:

(1) review of the specification of requirements;
(2) system review;
(3) functional unit evaluation;
(4) component analysis.

The checklist appears to be the principal means by which audits are presently carried out. A comprehensive list of items for use in the system and functional unit level audits can be compiled from the references reviewed:

safety	ergonomics
performance	cost/value
maintainability	stress tolerances
manufacture	environmental compatibility
standardization	fault finding
installation	adjustments
simplicity	support requirements
modular construction	reliability

The majority of this set of items, not in any order of merit, are commonly judged in a qualitative manner; a review team could discuss each point in turn and reach a consensus on the suitability of a design for some application. However, this cannot be accepted as a satisfactory approach to a decision process which effectively gives a 'stamp of approval' to the engineering aspects of a capital investment programme which may run into millions of pounds. Quantitative evaluations of design proposals should be made, especially in the current highly competitive industrial climate where there is a low rate of investment in British industry. Emphasis should be placed on those variables which relate specifically to maximizing availability and cost effectiveness over the life of a plant.

The composition of the design review team varies considerably in the reviewed publications. This is not surprising since the designs under review are different in character. The need for an independent chairman, designers of the system under examination, plus other designers are identified, but no further conclusions can be drawn from the literature with respect to the main team. A requirement for specialist expertise is stated, and this must be the case if quantitative assessments are to be made as suggested above.

The literature has established a four stage design review procedure as being generally applicable. However, there is no systematic approach to the design review in evidence which incorporates quantitative methods of design evaluation. Also there are no firm guidelines as to the composition of a team which would undertake a design review of production facilities. The review procedure and team composition are

Design review experience

inter-related since different demands will be made of the review team at different stages of the review.

Before these aspects can be considered in detail, the objective of a design review has to be established. It is clear from the literature considered that different authors have slightly differing perspectives of the design review objectives.

Chapter Two
Objectives

1 STATEMENT OF OBJECTIVES

The design review is much more than a detailed scrutiny following the original design in the manner that an examiner checks a student exercise. It must address primarily those problems which are most likely not to be, or are unable to be, fully taken into account during the design stage. One has to recognise that designers face a very complex task. Possible solutions to a problem have to be generated and then one alternative chosen to be worked through to a final design. With so much to consider, variables such as maintainability and reliability may not be fully taken into account when the main problem often appears to be that of obtaining a workable solution.

The design review may be defined as: *the quantitative examination of a proposed design to ensure that it is safe and has optimum performance with respect to maintainability, reliability and those performance variables needed to specify the plant equipment.*

Variables that specify plant equipment may refer to production rate, quality of assembly, etc., in a manufacturing plant, or to yield, ability to use cheaper feed stock, energy consumption, etc., in a process plant. The emphasis on maintainability and reliability does not mean that they should be dealt with in isolation. Wherever possible, all other factors have to be included. It is recognised that this may not always be possible, but the principle must be established within the definition. Also, availability is not specifically mentioned. It is a function of mean time between failures and mean repair time. The mean repair time consists of an actual or corrective repair time plus the time to organize manpower, spares provision, etc. The last items are not known at the design stage. Therefore, it is preferable to restrict the review to reliability and maintainability, where maintainability refers to the mean corrective maintenance time.

The safety of plant will have been considered throughout the design process. It would be presumptuous for the review team to insist on re-checking calculations. Mechanisms exist to ensure that certain facilities are as safe as can be expected: insurance companies have to be

Objectives

satisfied, and in the nuclear industry, the Nuclear Installations Inspectorate has to examine processes and certain handling techniques before a site licence is granted. The role of the design review team with respect to safety must be to formally record the fact that the appropriate internal and external authorities have been satisfied, and that the necessary safety standards have been adhered to (e.g., asbestos in air).

2 LEVELS OF DESIGN

A four level design review has been defined: review of the design specification, system review, functional unit evaluation, and component analysis.

The first level essentially consists of a dialogue between the eventual user and the design team (in-house or contractor). It is unrealistic to propose a formalized approach to this stage. The recommendation is that each point in the specification be discussed in turn in order to bring out any misunderstandings. The design team should be encouraged to ask questions at any stage if ambiguities or uncertainties arise. The term 'user' means those personnel who have responsibility for eventually running the plant plus those responsible for deciding the technical aspects of the specification. Three disciplines are thus identified for the review team at this stage: production, maintenance, and research and development.

The remaining three levels of the review correspond to the three main levels at which a design is undertaken when described generally **(12)**.

Design Level

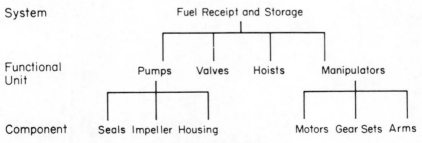

Fig. 1. Levels of design

However, there is usually some difficulty experienced in defining design levels in practice. A functional unit in one system may be a component in another larger system. Contradictions can also occur within one system. Consider the example shown in Fig. 1; pumps and motors appear on different levels, but really they are items of a similar kind with respect to size, composition etc. There is no simple answer to resolve this discrepancy, except to appeal to common sense when carrying out a review. The important point is not to let apparent inconsistencies disrupt the working of the review team, and to tackle each case sensibly in the manner which it deserves.

Chapter Three
System review

1 SYSTEM REVIEW PRIOR TO DETAIL DESIGN

A sensitivity analysis may be conducted concerning overall plant availability. Using flow charts and knowing the nominal production rates of various parts of a plant, including buffer capacities, it is possible to identify critical areas where a stoppage is likely to cause a total plant shutdown after a short time. It would not be possible to carry out a reliability analysis of the plant unless some very broad assumptions were made regarding the reliability of the many elements of the plant. However, a reliability analysis using conventional statistical methods may possibly be used to identify critical areas, but little confidence should be placed in the numerical results of the calculation, only relative values are significant.

These systems analyses would be expected to be carried out by any reasonably competent design team. The design review must not duplicate or trespass into this area. Instead, the role of the design review should be to ensure that the appropriate detail design teams are made fully aware of the existence of any critical areas. In this way, the design review complements the main design activity by ensuring that the necessary extra attention is given to certain aspects of parts of the main design.

Before proceeding with the detail design phase, the review team can take an overview of the proposed individual projects and comment on the relative risks associated with novel designs. No certainty of outcome is known. Also decision making under risk is a very doubtful proposition. The 'outcome' of a course of action is a function of many variables, and the probabilities of outcome for each variable cannot be estimated with acceptable accuracy. Decision making under uncertainty, where the probabilities of outcome are completely unknown, appears to be a possibility. Examples of the techniques for this kind of problem use a pay-off matrix, pay-off being with respect to one variable **(15)**. In reality, it is difficult to imagine a review team undertaking such an analysis and obtaining meaningful results. A more practical approach would be to discuss the design of 'high risk' projects with the team who will undertake the work. It will be quickly apparent if the proposal is unrealistic or if a lengthy research and development programme, before and after design, will be necessary. A recommendation not to proceed should be enforceable, or the need for a design procedure starting with operating requirements working towards a feasible solution strongly recommended.

2 SYSTEM REVIEW AFTER DETAIL DESIGN

Jones (**12**) defines the aim of systems engineering as: 'To achieve internal compatibility between the components of a system and external compatibility between a system and its environment'. He goes on to outline the design method as follows.

(1) Specify the inputs and outputs of the system.
(2) Specify a set of functions capable of transforming the inputs to the outputs.
(3) Select or design physical components capable of performing each function.
(4) Check the resulting assembly for internal and external compatibility.

Although couched in general terms, his description does fit engineering practice. Take the example of the system 'Fuel Receipt and Storage'. (Nuclear fuel reprocessing plant).

(1) The system boundary is defined by:
inputs – flasks containing irradiated fuel;
outputs – clean empty flasks; naked fuel elements for reprocessing; waste products.
(2) Necessary functions:
flask transport, manipulation, and cleaning;
pond storage;
fuel element handling and transport;
waste handling.
(3) Physical components:
hoists;
high pressure sprays;
waste pipeline, pumps, etc;
fuel element elevator;
master–slave manipulators.

System review

(4) Check that the contaminated liquid cleanser (waste) can be handled by the site waste treatment facilities, the throughput of the facility matches irradiated fuel generation from power station, fuel elements can be taken (after some storage time) by the decanning caves, etc.

Often, process plants are designed so that particular functions are carried out within individual buildings, or within clearly defined parts of larger structures. For example, a solvent extraction plant will be separated from a final product finishing line. The system boundaries used in the initial design should be determined by the inputs to, and outputs from, the separate processes. In this way, by ensuring compatibility within the domain of each building management, operating difficulties will be reduced since equipment will be properly matched. There is an opportunity for the design review to complement the initial design by defining systems for examination that encompass the whole production site rather than follow the initial concept. These system boundaries will cut across those of the original design and will ensure site uniformity. Alternative methods of system boundary definition are shown in Fig. 2. It is a practical proposition to take, say, remote handling equipment and to check through the various devices being proposed. To select one make of manipulator for one facility and a different make for another would not normally be sensible. Discussion between the review team and various designers can establish the reasons for the diverse choices. A sub-group may have to be formed to thoroughly analyse each manipulator to select the most suitable design. It is recognised that eventually different manufacturers' products may be selected if each has a machine suited to quite different specialised movements.

In general, 'horizontal' systems ensure site uniformity and standardization. Spares holding costs will be reduced and favourable purchasing agreements may be reached. 'Vertical' systems reduce problems within local working environments and should be used in the initial design exercise. Together, these two approaches should produce an optimum solution from both local and site viewpoints.

After the detail design stage, it is possible to carry out a reliability analysis for each 'vertical' system since details of the functional units comprising each system will be known. From a functional unit evaluation, which will be discussed later, more accurate reliability predictions are available. An interaction between the systems and

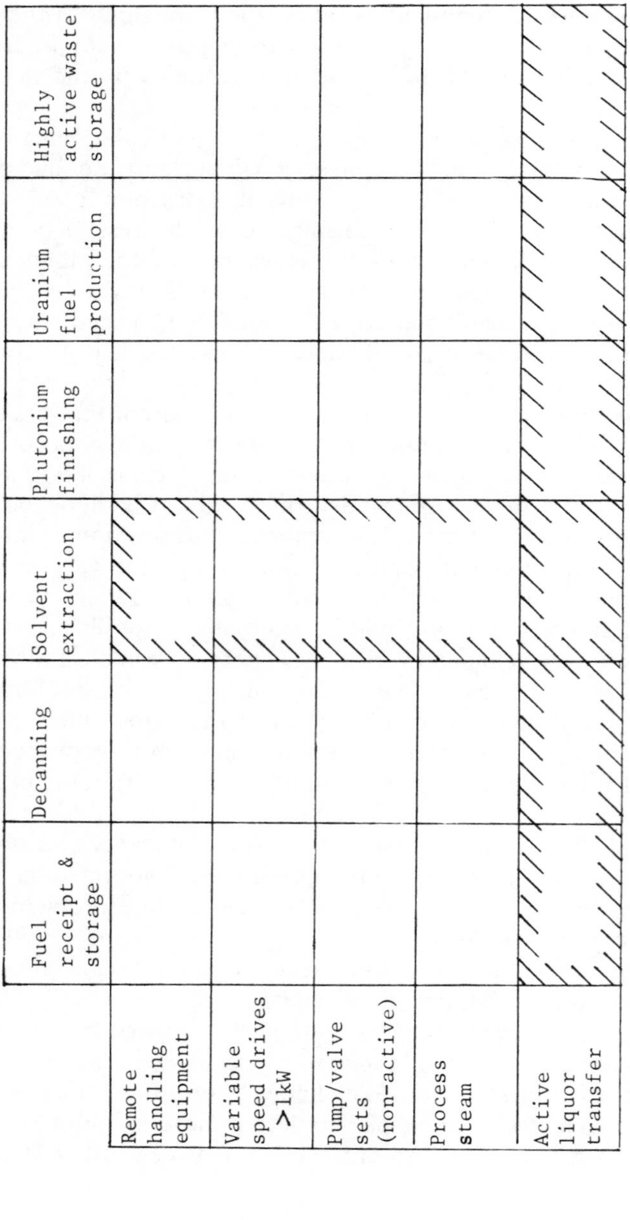

	Fuel receipt & storage	Decanning	Solvent extraction	Plutonium finishing	Uranium fuel production	Highly active waste storage
Remote handling equipment			////			////
Variable speed drives >1kW			////			////
Pump/valve sets (non-active)			////			////
Process steam						////
Active liquor transfer	////	////	////	////	////	////

Fig. 2. Alternative methods of system boundary definition

System review

functional unit levels of the review procedure has, therefore, been found. The interaction will be of an iterative kind. The functional unit evaluation is first carried out to predict reliability values from which the system reliability calculations can be made. The results of the system analysis may call for a re-design and evaluation of particular functional units. A revised system analysis can then take place and this process repeated for each system until a satisfactory solution has been found. The statistical methods for carrying out reliability analyses using event structures are well known **(16) (17)**.

Chapter Four
Functional unit evaluation

1 INTRODUCTION

The functional unit evaluation is a most important part of the design review. Units may be identified for detailed consideration as a result of: the general systems sensitivity analyses, from the study of 'across site' systems comprising functional units of a similar kind, from past bad experience with similar equipment, or at the request of a member of the design review team who is uncertain about the design of a novel machine. Projects initially designated as high risk ones should automatically be subjected to close scrutiny on completion of the detail design.

It is at the functional unit level that the engineer has the opportunity to embark upon a quantitative appraisal. Firm proposals have been set down and specific hardware proposed. The design review team should complement the original design work and not just duplicate some aspects of that work. The approach should be to 'stand back' from the design and assess the usefulness of the design proposal to the plant of which it is a part.

Most engineers will readily analyse a design proposal with respect to aspects of applied mechanics, thermodynamics, and manufacturing methods. This type of analysis typically leads to important conclusions relating to strength, efficiency, cost, etc. However, when alternative design proposals are analysed in this way, and all designs satisfy separate parts of a specification, but in varying degrees, then a meaningful method of comparison must be used to choose the 'best' design. The making of such a choice may lead to controversy in a design review team. The use of simple ranking methods do not appear to command much confidence. Commonly, a points score is directly attributed to the calculated result with respect to some variable, and the total score of a design is a measure of its overall usefulness. The probable drawback to this approach is the lack of precision in the determination of the points score.

A method is needed, therefore, that can quantitatively compare functional units, and in which confidence can be placed in any data not

Functional unit evaluation

generated by 'normal' engineering calculations. Such a method must not be so over-elaborate that it becomes unacceptable to practising engineers. However, it must be sophisticated enough to carry out its intended purpose satisfactorily.

2 UTILITY VALUE

A suitable design analysis method is one based on a set of value assessments for those variables considered to be most important. The derivation of the method is given in Appendix 2. Utility value is an appreciation of the usefulness of a device and may be related to a design variable as shown in Fig. 3. In this example, the graph refers to the carrying capacity of a robot arm at 'full stretch'. The usefulness or value of the machine, described on a utility scale of 0–10, would refer to a defined application. If the manipulator can handle 20 kg then it is as good as any other, but there is a pronounced fall in its value if the maximum load is less than 10 kg.

In general then, it is possible to derive a set of utility graphs corresponding to the variables to be included in a study. The analysis requires that the variables considered must refer to maintainability, reliability, and any performance variables needed to describe the equipment properly. In this way consideration is given to 'production' and 'technical' aspects of the design.

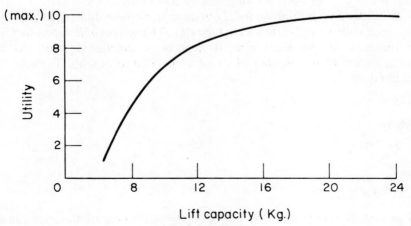

Fig. 3. Utility graph

Whilst a number of variables are considered appropriate for inclusion in the study, it may be that some are more important than others. This situation is accounted for by attributing scale factors or weightings to each variable, according to importance.

The derivation of utility graphs and scale factors is carried out with reference to the application alone, as illustrated in Appendix 3.

3 MACHINE ANALYSIS

The proposed equipment is analysed in a conventional manner, resulting in predictions of mean corrective maintenance time, mean time between failures, and those results pertaining to the other variables included. Any competent design analysis would yield such results. The strength of the method used here lies in the combination of the 'bare' findings in a meaningful form which takes into account the value of each of the predicted quantities to the proposed application. Quantitative maintainability and reliability predictions are discussed in Appendix 1. It is assumed that the reader is familiar with calculations with respect to stress analysis, fluid mechanics, etc.

4 SYSTEM UTILITY

Firstly, each calculated result is converted to a utility value by reading off the corresponding value from the appropriate utility graph. Secondly, the set of utility values obtained is combined with the scaling factors to give an overall rating: the *system utility*.

Consider a total utility score $\bar{U}$ obtained from individual utility scores, u_1, each with an associated scale factor k_1. The *system utility*, S, which is a measure of the total effectiveness of a functional unit and is independent of the number of variables needed to describe the unit, is defined as

$$S = \bar{U}/K$$

where

$$\bar{U}^{-1} = \Sigma\, k_1\, (u_1)^{-1}$$

and

$$K^{-1} = \Sigma\, k_1$$

The system utility rating is brought into the range of the individual utility graphs, say 0–10, using the factor K. For any unit described by n

Functional unit evaluation

variables, if each variable has a utility rating u then the system utility calculation must give a rating u also. It can be readily shown that the above definition gives this rating. Not all variables of interest need be included in the calculation. Lower bounds in performance may be prescribed with respect to some variables. This is akin to having a step utility graph from zero to maximum utility at the lower bound.

The system utility criterion may be used in several ways. All functional units comprising a system may be evaluated to identify a weak point in a system. This is possible since a simple rating, say on a scale of 0–10, is obtained even when different types and numbers of variables are used to describe the functional units. If alternative designs are being considered for one particular application, the system utility criterion provides a quantitative method of evaluation. If all competing designs meet all parts of a specification, but in varying degrees, then it is especially useful to have such a criterion.

The method should not be considered as a straightforward evaluation exercise only. The concept of system utility can be used as an optimization tool also. In this way, it is possible to trade-off one variable's performance with another to achieve a maximum usefulness. It may be that a specification may be relaxed in one respect to obtain more valuable gains another way. The case study given in Appendix 3 illustrates this point. In that example a machine's throughput is kept to a low, but acceptable, level in order to improve reliability considerably.

5 GENERAL CONSIDERATIONS

Of course it is not always essential to undertake a comprehensive functional unit analysis using the system utility concept. Such a formal method is suited to novel designs or for particularly critical plant items. In many cases, a less stringent approach is more appropriate, and indeed, other factors have to be considered in a design review which cannot be included in the system utility calculations.

The operation and maintenance practices for equipment should be reviewed. Typical questions that should be asked are as follows.

(1) What skill levels are required to operate equipment?
(2) How much routine maintenance and fault finding can operators carry out independently from maintenance departments?
(3) Will demarcation problems arise during operation and maintenance? This may be a significant problem if modular construction

methods have been used where mechanical, electrical and instrument connections are simultaneously made in one simple operation to 'ease' maintenance.
(4) Has sufficient attention been paid to fault finding techniques and is condition based maintenance desirable?
(5) Is there any obvious excessive spares handling requirement?
(6) Have the suppliers of equipment a good reputation for support (technical and spares provision) once equipment is operational?
(7) Should a maintenance contract be negotiated, say in the case of equipment with novel features where there is a high risk factor with respect to reliability?

The above list is not exhaustive, but the questions posed indicate the kind of comprehensive qualitative analysis that should always take place. In addition, the check list given in section 2 of Chapter Five for component analysis may equally be applied to functional units if the system utility method is inappropriate.

It can be seen, therefore, that the skill requirements of the design review team have to be broad based.

Chapter Five
Component analysis

1 SCOPE OF ANALYSIS

The demarcation between functional unit evaluation and component analysis is not precise, and has been discussed in Chapter Two, section 2. The question is one of scale. For our purposes, a functional unit will be considered to be a machine, such as a centrifuge or, as discussed in the case study (Appendix 3), a piece of equipment for assembling battery plates and separator in a particular sequence. A component is taken to be a device such as a motor or pump.

It is clearly impractical to consider a general site survey of components. Some cases will naturally follow from analysis at the functional unit level. Certain component classes may be identified for detailed analysis where there are particularly demanding duties, for example, the seals in a complete process line carrying a corrosive fluid.

2 ANALYSIS METHODS

2.1 General
A check list is a most useful tool with which to determine whether as much thought as can be reasonably expected has been given to component selection. Below is a list of items for consideration. Each will take on different importance according to the application in question, hence they are listed in alphabetical order.

Accessibility
Adjustments
Cost
Environmental compatibility
Ergonomics
Fault finding
Installation
Maintainability
Modular construction
Quality assurance
Reliability

Sensitivity to feed stock or material quality
Simplicity
Standardization

It need hardly be said that *safety* is a major concern.

2.2 Components handling fluids
Particular problems arise when components are used for fluid handling. Corrosion resistance is not the only problem for, if particulate matter is carried in the fluid even in very small quantities, then there is a possibility of seizure of moving parts if solids are deposited. A useful procedure for such cases is as follows.

(1) List, in sets, all mating parts that have relative motion, either during operation or in maintenance operations, then check their sensitivity to seizure.
(2) Examine the internal contours of the component in the attitude of installation (not simply as drawn) for regions of fluid hold-up. Note if stagnant zones may be formed under conditions of fluid flow. Flushing with a cleansing agent will not clean the component if stagnant zones are present.

Fluid traps may also cause problems of corrosion where free-flowing process fluid does not. When a mixture of two or more fluids is handled the potentially harmful effect of one fluid may be inhibited by the action of others mixed with it. However, if the component is non-draining or if stagnant zones are formed during operation, then there is a possibility of fluid separation occurring. A concentrated layer of liquid may then corrode parts of the component that would otherwise be resistant to the action of the fluid mix. A trap in itself need not necessarily be prohibitive if corrosive liquids are handled. But if the liquid will be held close to seals, screw threads, or highly stressed components, then in-service problems will ensue.

2.3 Reliability
It is rarely possible to apply reliability theory to component analysis since relevant data is commonly not generally available. However, a compromise between formal analytical methods and a reliance on personal experience judgements is possible.

(1) Examine the design and list all the events that may lead to failure.

Component analysis

(2) Analyse the event structure using conventional statistical methods letting each event have a probability P_i which is unquantified.
(3) Repeat this process for alternative designs and compare the results.

No requirement has been made for comprehensive statistical data. Instead, a comparison between different designs is made on a relative basis. When a different component can be specified to perform the same function, e.g., bellows sealing rather than stem packing in a valve, then personal judgement must be used to compare the relative merits of each method. The survival probability of one component is adjusted to a fraction of its counterpart in the competing design, and a comparison of the design is again possible.

For component reliability then, the review team has to rely on a semi-qualitative comparison between alternative designs, and on such quantitative data as is generally available.

3 APPLICATION OF THE ANALYSIS METHODS

The detailed component analysis method described above cannot be undertaken for all cases. Therefore, the skill of the design review team must be to identify subjects for this kind of analysis. Appendix 4 gives two examples of how the component analysis method is used.

The above approach to component analysis eliminates many practical problems that can arise during plant use. Attention is directed to key areas, and forces analysts to seek advice on matters beyond their present knowledge. As in the case of functional unit evaluation, the task of undertaking a component analysis would be delegated. The method by which it is carried out should be laid down by the review team, and a report of the findings required.

Chapter Six
The structured design review

1 PROCEDURE

The design review is defined as: *the quantitative examination of a proposed design to ensure that it is safe and has optimum performance with respect to maintainability, reliability, and those performance variables needed to specify the plant equipment.*

The review is carried out with reference to different levels in a design, and its activities span the design process from a review of the design specification to component checking before the issue of drawings for manufacture. The review complements the work of the design teams rather than being a duplication exercise as a 'check on calculations'. The procedure, as described above, can be summarized as follows.

I   *Activity*   Review of the design specification (requirements).
     Purpose   To ensure that the significance of all the points contained within the design specification are understood.
     Timing   Prior to the commencement of any design activity.

II   (i) *Activity*   Systems level review.
         Purpose   To identify critical areas of the design that may affect plant availability, and to communicate to the detail design teams the necessity to pay particular attention to these areas. Comment on the advisability of pursuing projects with a high risk content.
         Timing   Prior to the start of detail design.
     (ii) *Activity*   Systems level review.
         Purpose   Examination of systems formed across the plant to ensure uniformity and compatibility. To maximize the reliability of systems formed by process considerations.
         Timing   After the completion of the first detail designs.

III   *Activity*   Functional unit evaluation.
     Purpose   To maximize the usefulness of a functional unit to the system of which it is a part.
     Timing   After the completion of the first detail designs.

IV *Activity* Component analysis.
Purpose To check that certain important sets of components will fulfil their required duty.
Timing After the completion of the first detail designs.

It is vital that sufficient time is allocated for the design review when planning a project programme. Sufficient emphasis should be placed on the review in the early stages of design, for it is wasteful to spend large amounts of time and money correcting mistakes which arise because of poor basic design principles or lack of appreciation of the business need. Good foresight and a sound design foundation will undoubtably save money in the long term.

There is no reason to presume that the use of a critical systematic design review will lengthen the time taken to achieve a working plant. A shift in the time taken for particular activities will, however, be apparent. The longer design time should be (more than) compensated for by a reduction in commissioning time. A further payoff will also be in the form of good plant availability and performance, particularly in the initial production phase, but, importantly, throughout the plant life.

2 TEAM COMPOSITION

Prior to the detail design stage, the role of the review team is to ensure that the initial design specifications are fully understood, and to comment on and to communicate the results of calculations regarding plant sensitivity with respect to availability. After the detail design stage, the work of the review team is basically to identify certain items for further evaluation. The detailed functional unit evaluations and component analyses would be expected to be delegated. However, although the review team does not have to actually carry out specific analysis exercises, the team members must understand the principles of the methods of analysis and their limitations. In making recommendations which necessarily involve extra work, the members must also have sufficient authority within the company to ensure that the work will be carried out.

In order to carry out a design audit effectively the team must be multidisciplinary in character. Also, specialist expertise will be needed to undertake detailed studies. Taking into account all the above discussions and the points raised from the literature reviewed, the following team composition is suggested.

(i) *Design* (A) There must be a direct link with the original design team in order that comments may be made on the implications and feasibility of suggested modifications.
Strength on team: 2 (1 only needed prior to detail design stage).

(ii) *Design* (B) A designer with relevant experience, but not involved in the design under examination, should be available to advise on the approach being taken to design activities.
Strength on team: 1 (needed only after detail design stage).

(iii) *Research and development* Specialist expertise will be necessary to give advice in certain areas (e.g., corrosion science, vibration analysis). A person with broad experience is required who can recognize the need to co-opt a specialist or to commission detailed studies as necessary (e.g., stress analysis of non-standard pressure vessel connections).
Strength on team: 1 (plus co-opted members as required.

(iv) *Maintenance* The original design team should have had feedback of maintenance field data. However, a maintenance engineer on the review team will provide first hand experience and should be capable of identifying problems associated with feasibility of operation, staffing levels, demarcation, etc.
Strength on team: 1.

(v) *Production management* Experience of similar plant is always useful. Staffing levels, industrial relations, and management efficiency in the new plant are relevant.
Strength on team: 1 (during specification review and after detail design only).

(vi) *Safety* Problems associated with personnel and environmental safety are always of paramount importance.

The structured design review 31

>Any new plant design should be vetted with respect to these conditions, no matter how innocuous the process materials may appear to be.
>*Strength on team:* 1 (presence not always required, but should see the proposals at every stage).

(vii) *Project engineer*   The contribution of the project engineer is very important. He is the link between design, commissioning, and production. The project engineer also will be familiar with most aspects of the design since he is responsible for the transition from drawing board to reality. The project engineer, therefore, is in the best position to act as chairman of the design review team.
>*Strength on team:* 1.

The number on the review team is between six and nine depending upon the topic under discussion, which is a manageable group.

The question of whether a design review should be undertaken by personnel outside the eventual operating company must be addressed. No doubt there is an argument that 'fresh minds' should bear on the subject and, therefore, an 'outside team' should be employed. However, the strongest argument surely is that the personnel who will be ultimately responsible for operating the plant should be heavily involved in the reviewing procedure. After all, their livelihoods will depend upon the plant long after the review team has been disbanded. Also, the presence of the extra designer not connected with the design under examination provides a fresh mind. These arguments weigh in favour of an 'in house' review team, but a compromise situation is possible whereby the extra designer and the research and development expertise is provided by a specialist organization. The latter contribution could only be justified if the operating company did not already support a research and development department of its own.

3 MANAGEMENT AND CONTROL

A four part design review procedure has been proposed which is carried out with reference to different levels in a design. A multi-diciplinary team of about seven has been identified to carry out the review and they should possess a wide range of experience and expertise. Team experience is vital since it is from knowledge of past difficulties that certain subjects are identified for further evaluation. This is a form of information feedback.

The review team effectively manages the analysis of a plant design from the review of the design specification through to detailed component checks. It is the responsibility of the review team to ensure that, within reason, potentially serious operating difficulties with the plant are avoided. The responsibility for decisions concerning the acceptance of design proposals which have been subjected to particular examination by the design review team rests with the team.

Specific analysis techniques which are applied to systems, functional units, and components form integral parts of the review procedure. These analyses are best carried out by personnel not part of the review team but who must of course be present when the results are discussed. There are two main reasons for this delegation of work. The person with broad experience of plant operations may be reluctant to undertake detailed analyses, especially if they involve mathematics. More importantly, by delegating such duties the review team can retain its overall control of the design analysis without becoming submerged in detailed calculations. The review team must, however, understand the principles of an analysis method in order to properly interpret the findings.

Experience within the review team is important. For the case of the system review prior to detail design, the team has to comment on the relative risks associated with any novel designs. Attitude to risk is dependent upon the operating environment of the company – both in the workplace and the business climate and, therefore, experience within the company is required. After detail design, two types of system have been proposed for analysis: one has boundaries defined by the processes carried out, and one groups particular kinds of equipment irrespective of their process applications. To undertake this boundary definition exercise effectively, the review team must have a good understanding of the industry. Therefore, it is unlikely that a review could be satisfactorily carried out under contract.

Functional unit evaluation is a major part of the design review. At this stage it is possible to see and comment on the exact form of the equipment proposed for a new plant. The system utility criterion is ideally suited to handle quantitatively the analysis of functional units with respect to a number of variables. Generally, it would not be possible to undertake this kind of assessment during design synthesis because there are too many problems to consider. The method is,

The structured design review

therefore, particularly suited to an evaluation exercise in a design review. The review team can set a minimum level of system utility rating for all plant equipment. Any unit which has a system utility score below this level would then be rejected.

Firstly, the review team specifies the variables needed to describe a functional unit. Using their experience, the members of the team then make the value judgements for each variable. A sub-group can subsequently proceed with the detailed analysis, reporting back its findings to the review team. The final decision concerning the acceptance of a functional unit is taken by the review team.

Thus although specific detailed analyses may be commissioned by a review team at system, functional unit, or component levels, the overall control of the design review exercise and decision making responsibility is retained by the review team.

Appendix 1

MAINTAINABILITY PREDICTION

Consider a machine with no case history but made from components which are not novel. A machine failure is assumed to occur if a component fails. If there is built-in redundancy, then the multiple component sets can be reduced to an equivalent component which will cause a machine failure if that component fails. The machine then consists of a set of components, the failure of any one giving rise to a machine failure. These are a set of primary modes of failure. Secondary failures may occur as a result of the primary failure.

For each primary mode of failure, it is possible to estimate a corrective repair time. Repair may be done by replacing a component with a new one, or by removal and effecting a bench repair before replacement. In the former case it has to be assumed that a replacement is immediately available. Spares provision is beyond the control of the designer, so a time to obtain spares cannot be included. A bench repair time can be estimated and included in the corrective repair time for each failure mode. For cases where a secondary failure will occur, then the maintenance time for that component has to be included in the corrective maintenance time for the primary failure mode. It may happen that a component repair time is included twice, once as a primary mode failure in its own right and secondly as part of a total repair time following another primary mode failure. To estimate repair times, a complete set of assembly drawings has to be available. Under special circumstances, some hardware may be available. The assistance of an experienced tradesman is highly desirable when repair times are estimated, and the working conditions must be precisely specified to eliminate the possibility of misunderstandings.

The probability of failure for each primary failure mode may be estimated from tabulated failure rate data **(20) (21)**. Such data refer to components operating under 'standard' conditions. The failure rate for a component λ_i working under other conditions is obtained by multiplying the basic failure rate by three service factors, Z, pertaining to: variations in nominal rating, temperature, and other environmental conditions.

Appendix 1

$$\lambda_i = Z_1 . Z_2 . Z_3 \, \lambda_{basic}$$

Whilst the failure rates obtained in this manner are not very accurate, the technique is the best quantitative one for which data is generally available. The probability of failure for each primary failure mode is given by

$$F_i = 1 - e^{-\lambda_i t}$$

where t is the design life of the equipment. This equation assumes that each failure rate is constant and only the 'useful life' of the machine is being considered. It is difficult to justify the consideration of the 'running in' and 'wear out' phases when the failure rate is not constant. The accuracy of prediction does not warrant the increased mathematical complexity.

One may consider the mean corrective repair time of the machine $\bar{M}$ as a variable which may take the value of any of the estimated repair times, M_i, caused by a primary failure with probability of occurrence, F_i. Given that a failure has occurred, the probability that the variable will take the value M_i is

$$\frac{F_i}{\Sigma F_i}$$

assuming that all the primary modes of failure are independent. The expected value of the variable is, therefore

$$\bar{M} = \frac{\Sigma F_i M_i}{\Sigma F_i}$$

The above equation gives an estimate of the average time that it will take to repair the machine, under prescribed conditions, after a failure over the life of the equipment. It is interesting to note that mean corrective maintenance time is a function of failure rate and the design life.

Repair times for each component are best decided by reference to operations carried out under workshop conditions. An allowance can then be made for field conditions, and the expected mean corrective maintenance time adjusted accordingly.

An expression for the mean corrective maintenance time has been suggested by Smith (3).

$$\bar{M} = \frac{\Sigma \lambda_i M_i}{\Sigma \lambda_i}$$

No supporting analysis is given. The two expressions above for $\bar{M}$ can be shown to be equivalent under certain conditions.

$$F = 1 - e^{-\lambda t}$$
$$= 1 - \left\{1 + (-\lambda t) + \frac{(-\lambda t)^2}{2!} + \frac{(-\lambda t)^3}{3!} + \ldots\right\}$$
$$= \lambda t$$

provided λt is small

Then

$$\frac{\Sigma F_i M_i}{\Sigma F_i} = \frac{\Sigma \lambda_i M_i}{\Sigma \lambda_i}$$

If there is a large number of components and the design life is long, then λt may not be small. However, the alternating sign of the higher order terms of the expansion will probably reduce their contribution.

It is interesting to note that reliability and maintainability are not independent at the functional unit level.

Reliability prediction may be carried out using failure rate data obtained as described above. Again it is only sensible to consider the 'useful life' period where the failure rate is approximately constant. When data are not available for the particular component required, then an estimate has to be made using the nearest case available.

It is worth noting that Quirk (**22**) describes a method for obtaining a reliability index for a machine or device. This is not a survival probability as conventionally defined, but it is a ranking method for comparing alternative designs. A failure probability is required to estimate the mean corrective maintenance time, so Quirk's Logic Design Factor method will not be discussed in detail. However, the method is based upon subjective assessments of the reliability of components relative to each other on an arbitrary scale of one to five. Importantly, Quirk has shown that the inclusion of such relative subjective assessments does work satisfactorily.

Appendix 2

SYSTEM UTILITY

1 Introduction

System utility is a means by which functional units may be compared in a selection exercise, or by which a weakness may be identified in a process or production line. It may also be used as an optimization tool to maximise the usefulness of a machine to the system of which it is a part.

Difficulties arise when comparing competing designs if all satisfy a specification, but to varying degrees. The question of 'Which one is best?' is a natural question, and one which conventional engineering analysis, using the principles of mechanics, etc., cannot answer. Such analysis alone is also unable to determine which part of a plant will under-perform with respect to the rest of the facilities, unless a distinct cause of failure is identified. Any attempt to optimize a design by relaxing part of a specification to obtain 'better' gains elsewhere also needs more than conventional analysis. That is not to say that the principles of mechanics, thermodynamics, etc., are not required. Rather a requirement for more information and further analysis of that information to be able to carry out the tasks described above is identified.

2 System effectiveness

The concept of system effectiveness has been proposed as a measure of the degree to which equipment approaches its inherent capability and achieves ease of maintenance and operation **(10)**. It is defined as

    Performance × Reliability × Availability

and assumes that these factors are independent. However, it can be readily shown that steady state availability has a simple relationship between reliability and maintainability, given by

$$A = \{1 + \lambda(M)\}^{-1}$$

where λ is the instantaneous failure rate and M is the mean repair time. Although the assumption of independence does not strictly hold, the concept is nevertheless useful. A modified definition may be written as

    Performance × Reliability × Maintainability

The units in which system effectiveness is measured, and the way in which the concept is applied, are not dealt with in the above reference, it is most important that these two points be considered. The formal definitions of reliability and maintainability, respectively, refer to the probability that an element will not fail to perform its required duty under specified conditions within a given period, and the probability that a device can be maintained by specified practices within a certain time. Reliability and maintainability are, therefore, dimensionless and, hence, system effectiveness has the units in which performance is measured. However, the question of 'what is a machine's performance?' must be asked. For example, in a drying furnace is the moisture content of the processed material a measure of performance, or is the volume (or mass) throughput the correct measure? Obvious problems occur if more than one variable is needed to describe performance. The inclusion of maintainability and reliability is established by the system effectiveness concept. Clearly then, a common scale of measurement for different design variables is needed.

3 Utility

Recourse to value theory provides a common method of measurement of dependent design variables such as maintainability and moisture content. Utility value is an appreciation of the usefulness of a device and may be related to design variables as shown in Fig. 4. This graph refers

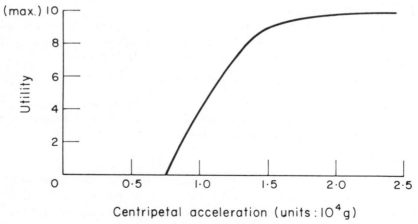

Fig. 4. Centrifuge utility graph

Appendix 2

to a centrifuge application and describes the usefulness of a centrifuge for separating particles (<5μ) of titanium carried in suspension. An arbitrary scale of 1–10 has been used for the utility axis.

There is a pronounced fall in the value of a centrifuge if it generates a centripetal acceleration of less than 1.5×10^4g. Below this level the titanium will not be extracted, therefore the machine has little value. Separate utility graphs can be drawn for any set of variables that describe a functional unit. Maintainability and reliability must be included in that set. Maintainability and reliability are best described, respectively, by mean corrective maintenance time and mean time between failure when deriving utility graphs.

The relative importance of variables may be taken into account by the application of scaling factors. If maintainability and reliability were considered equally important and twice as important as acceleration then scale factors of 2:2:1 would be applied to the respective utility graphs. If the number of variables considered is small then simple reasoning will suffice. For cases where the number of variables to be considered is large, or where it is required to demonstrate formally how the scaling factors are derived, then the interactive matrix method can be used. This is a useful method to permit a systematic search for relationships between (or relative importance of) variables **(12)**.

In general, then, it is possible to sketch out utility graphs for those performance variables that are considered to be important, and for mean corrective repair time and mean time between failure. A calculation of system effectiveness may be made by reading off from the appropriate utility graph the score of a design proposal with respect to each variable, and multiplying the scores together to give the numerical value of system effectiveness. There are two disadvantages to this. Scaling factors cannot be used since the multiplication process just multiplies the factors together and does not take into account the relative importance of the variables. Also, the calculated value of system effectiveness (SE) is very sensitive to a change in a utility (u_i). From simple differentiation

$$\frac{\partial (SE)}{SE} = \frac{\partial u_i}{u_i} \text{ for any variable } i$$

This is particularly important when a utility graph has a steep gradient. A small error in the calculated value of a variable will result in a large change in the utility value. The percentage change in the calculated value of system effectiveness will be equal to the percentage change in any of the variables considered.

Alternatively, we may look at other ways of combining utility scores. Instead of multiplication, the total utility (U) for i variables may be defined as

$$U = \sum_{\text{all } i} u_i$$

Scale factors may be used to modify each utility score (by multiplication) and the total utility is clearly much less sensitive to changes in individual utility values. However, summing individual utility scores may hide a very low score with respect to one variable if several other high scores are obtained. This is not satisfactory since a functional unit may be evaluated as good using this method whilst being virtually unmaintainable, which is contradictory to the aims of a design review.

An attractive method of combining utilities, without reference to scale factors, has been suggested by Siddall (**18**). The total utility is given by

$$U^{-1} = \sum_{\text{all } i} u_i^{-1}$$

The numerator of the expanded expression for U is formed by simple multiplication of the individual utility values which readily identifies a very poor performance with respect to one parameter. Taking the example of three variables

$$\frac{\partial U}{U} = \frac{\partial u_1}{u_1} \left\{ \frac{u_2 u_3}{(u_1 u_2 + u_2 u_3 + u_1 u_3)} \right\}$$

therefore the total utility is less sensitive to changes in individual utility values. The inverse method appears to combine the better attributes of the addition and multiplication methods.

It was suggested that there was insufficient confidence in the simple points method for ranking alternatives due to a lack of precision. The advantage of a utility graph is that attention can be focused onto a single variable, independently of the machine design and other variables, resulting in a formulation of a good assessment of value. The utility score with respect to any variable then becomes a systematic process based on conventional mechanical engineering calculations.

Appendix 2

Clearly, the set of utility scores contains extra information to that contained within the raw calculations: a quantitative assessment of usefulness. The utility graphs are not just a means of non-dimensionalizing quantities.

4 System utility

The performance variables which describe different functional units will vary and by the methods described above the total utility value calculated for a unit depends upon the number of variables considered. This is not important when comparing alternative designs for one application, but a comparison of units having separate and different duties is not possible. There is a good reason for wanting to compare different units. A standard measure applied across the set of functional units which form a system will identify any weakness in that system. Availability is suitable for this purpose with respect to production losses since reliability and maintainability are both applicable to all parts of a system. However, availability is incomplete as a measure of the total contribution of a functional unit to a system because performance parameters are omitted. The utility combinations described above do not immediately lend themselves to such use due to their dependence upon the number of variables considered.

Consider a total utility score $\bar{U}$ obtained from individual utility scores, u_i, each with an associated scale factor, k_i. A measure of the total value of a functional unit, system utility S, which is independent of the number of variables needed to describe the unit may be defined as **(19)**

$$S = \frac{\bar{U}}{K}$$

where

$$\bar{U}^{-1} = \Sigma \, k_i \, (u_i)^{-1}$$

and

$$K^{-1} = \Sigma \, k_i$$

The system utility rating is brought into the range of the individual utility graphs, say 0–10, using the factor K. For any unit described by n variables, if each variable has a utility rating u, then the system utility calculation must give a rating u also. It can be readily shown that the above definition gives this rating. Not all variables of interest would be included in the calculation. Lower bounds in performance may be prescribed with respect to some variables. This is akin to having a step utility graph from zero to maximum utility at the lower bound. Maintainability and reliability should always be included within the system utility calculation, plus those performance variables deemed necessary in each particular case.

The system utility definition clearly embraces the concept of system effectiveness, but in a form readily applied in a quantitative manner in a design review exercise. The system utility of any functional unit may be calculated and compared with either an alternative proposal for that duty or with another unit performing a different duty. The strength of the method lies in the fact that a review team is able to set a minimum system utility rating for a plant against which any functional unit may be assessed. This minimum rating would be set prior to the detail design stage.

In most engineering practice, value considerations and subjective judgements are made regularly, but usually not in a structured manner. The formalized procedure proposed seeks to structure the process as an aid to decision making. Careful and considered professional judgement on the practical aspects of the study, such as which performance variables to include, is essential for the method to work.

Appendix 3

CASE STUDY: A DESIGN REVIEW OF AN ASSEMBLY MACHINE FOR BATTERY COMPONENTS*

1 Project description

This case study is concerned with a machine which assembles a set of battery plates prior to their insertion into a battery case. Plate assembly machines of various designs are currently used in production plants. However, none has developed a good reputation, particularly with respect to reliability. It was decided by Chloride Technical Ltd that a new generation of such machines was needed for their organization's production facilities, and that the design and development work should be undertaken 'in house'. As part of the development programme, a design review was considered necessary.

The equipment is referred to as a 'wrap–stack' machine, and its function is as follows. Separate batches of positive and negative battery plates (150 × 150 × 3 mm approx.) are supplied to the machine, along with a plate separator material. The separator is in strip form of approximately the same thickness and width as a battery plate and is fed from a roll approximately 1 m diameter.

A plate is taken and a length of separator is automatically fed, cut to length, and wrapped around the plate. A plate of the opposite kind is then taken and stacked on top of the wrapped plate. This process is repeated until the requisite number of plates has been stacked. The assembly of alternate positive and negative plates interspaced with one layer of separator material is then transferred from the stacking area ready to be packed into a battery case. Stacked plate sets have to be precisely and accurately assembled since mis-alignment gives rise to poor battery performance; the production rate is also important. The battery manufacturing rate is currently limited by this 'wrap–stack' machine; other parts of the production process have a better performance in this respect.

A general view of the machine is given in Fig. 5(a) with a close-up of some details in Fig. 5(b). The height of the machine above ground

---

*This case study was originally published in *Maintenance Management International*, Volume 4 (1984), pp. 237–249, under the title 'A design audit case study at functional unit level', and is reproduced here courtesy of Elsevier Science Publishers, Amsterdam.

Fig. 5. The 'wrap–stack' machine: (a) general view, (b) some details

Appendix 3

(excluding the roll of separator material) is approximately 1.5 m. Details of the components from which the machine is constructed are given in Table 1.

Table 1. Components, failure rates, and repair times

Primary drive unit	No. off n	λ (per 10^6 hr)	Corrective repair time M_i (hr)
(a) DC Motor (0.5 hp, 2000/73 revs/min)	1	10	0.5
(b) Spur gear (steel)	2	10	2.0
(c) Torque limiter	1	3	1.0
(d) Flexible coupling (polymeric material)	1	10	1.5
(e) Bevel gear box	2	65	0.75
(f) Microswitch (lever-cam operated)	6	0.5	1.0
(g) Keyed rigid coupling	1	0.4	1.5
Shuttle drives			
(a) Rolling element bearing	2	15	0.5
(b) Pivot	12	1	0.5
(c) Adjustment screw	3	0.02	0.5
Separator feed drive			
(a) Toothed belt	2	40	0.1
(b) Rolling element bearing	2	15	0.5
(c) Timing mechanism (light mechanical assembly, a bought-out item with no details provided)	1	11	1.5
(d) Belt pulley (keyed on)	3	0.2	0.5
(e) Knurled roller: (i) rolling element brg.	2	15	2.0
(ii) coil spring	4	1	2.0
(iii) knife	1	50	2.0
(f) Idler roller: rolling element brg.	2	15	1.5
Shuttle carriages			
(a) Linear bearings	8	20	2.5
(b) Torsion spring (for pawls)	2	0.2	0.75
(c) Self-lubricated bushes (plate grasp)	2	5	1.0
(d) Compression springs: (i) plate grasp	1	1	1.0
(ii) plate adjust	1	0.2	0.5

Plate feeders (proprietary design)

(a)	Air cylinders		10	3	1.0
(b)	Lubricated bush		26	5	1.0
(c)	Pivot		2	1	1.0
(d)	Pin/reamed hole		18	15	1.0
(e)	(i) Brass bushes	(plate stop, one	2	5	1.0
	(ii) Tension spring	side only)	1	0.2	1.0

Stacked plates conveyor

(a)	Motor gear unit (0.25kw, 1400 revs/min motor)	1	10	0.3
(b)	Chain	1	1	0.25
(c)	Chain wheel	3	0.2	0.25
(d)	Rolling element bearing	4	15	0.75
(e)	Conveyor belt (polymer, endless)	1	200	1.5

Pneumatic controls

(a)	Solenoid operated 'on–off' valves: solenoid	} 6	30 {	0.5
	valve			0.5
(b)	Controller	1	300	1.0
(c)	Junction box	1	5	0.5

Stacked plates handling

(a)	Chain		2	1	2.5
(b)	Chain wheel (keyed to shaft)		4	0.2	2.5
(c)	Toothed belt		2	40	2.5
(d)	Belt pulley (keyed on)		4	0.2	2.5
(e)	Rolling element bearings		8	15	2.5
(f)	Width adjustment screw		1	0.02	2.5
(g)	Stepping motor drive		1	5	2.5
(h)	Spur gear: (i) nylon		1	100	2.5
	(ii) steel		1	10	2.5

2 Utility graphs and scale factors

2.1 Method of study

The first step was to establish a set of utility graphs for use in the evaluation exercise. This was carried out with reference to the eventual production conditions (dirty environment, not highly skilled labour) and the type of machine in question (one with many moving parts). No reference was made to any specific details of the machine being

Appendix 3

designed. Therefore, there was no possibility of 'fixing' utility graphs to suit the machine.

Five members of the company's staff were interviewed independently. Collectively they had experience of previous 'wrap–stack' machines or similar equipment, plant commissioning, prototype design and development, and design-draughting. Preferably, some personnel from production operations should have been included, but this was not possible. However, several members interviewed had worked with production staff and the study had to rely on their experience.

Each person was asked to give a utility score, out of ten, for a given performance of a hypothetical machine with respect to each design variable being studied. The extremes of performance scoring 0 to 10 were first established. Then the intermediate ranges were quantified by working alternatively from each boundary. Any discontinuity at the meeting at mid-range was smoothed by random questioning throughout the centre range. In many cases it was common for the initial assessment to change in a repeat question. However, on subsequent examination across the mid-range of the variable, the utility values tended to settle down.

Attention was only turned to the next variable when the interviewee was giving consistent results and appeared satisfied with his value judgements.

2.2 Variables considered

It was stated in Chapter Four that maintainability and reliability should both be included in the study, plus those performance variables that are needed to describe the machine. Two performance variables were initially identified: stacked plate alignment and machine output. It was decided, however, that plate alignment really had a step utility graph. Either the stacked plates were to a standard which was acceptable or they were rejected. All plates stacked better than or equal to the minimum acceptable standard were of equal value. All below that standard had zero value since their correction involved a manual adjustment with practically a uniform cost. Therefore, the only performance parameter considered in the utility study was machine output. Questions were put to interviewees on the following variables in the order given.

(1) *Availability*

Figure 6 shows the data obtained. A reasonable correlation was

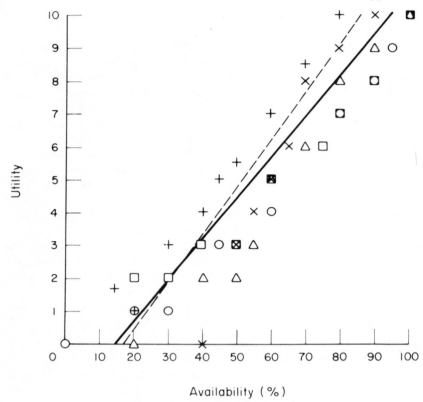

Fig. 6. Utility graph: availability.
(Note: each symbol in Figs. 6–9 refers to data generated by a particular individual)

obtained probably due to the fact that availability has a defined range of 0–100 per cent and most engineers have a 'feel' for this variable. The broken line was derived using data from two sources only. This was used to calculate a mean time to repair curve which is discussed below. Availability was not included in the system utility calculations since it contributes no additional information if reliability and maintainability are included.

Appendix 3

(2) *Mean time between failures (MTBF)*

Only two sets of data fell within what was considered to be a feasible range: up to 100 hours, a level of achievement which had never been exceeded with this type of equipment. Two sets of data are given in Fig. 7; the agreement is good. The graph represents a lenient view since, with one exception, the remaining data fell far to the right of that plotted. The remaining data did not contain a utility value greater than 3 for an MTBF of less than 300 hours. The two engineers whose data are plotted in Fig. 7 have substantial field experience.

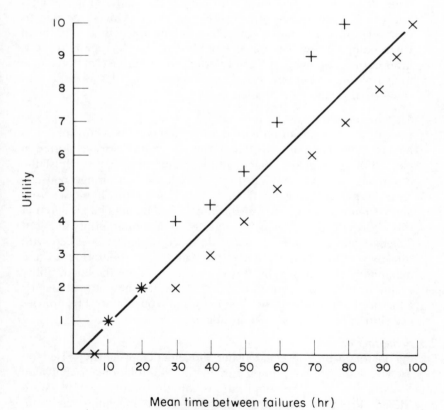

Mean time between failures (hr)

Fig. 7. Utility graph: mean time between failures

(3) *Mean corrective maintenance time* ($\bar{M}$)

Figure 8 shows both a mean time to repair curve and the data points for mean corrective maintenance time. The former was calculated from the MTBF graph (Fig. 7) and an availability graph using data only from the same sources (shown as a broken line in Fig. 6). It can be seen that $\bar{M}$ is small compared to mean time to repair for any utility value. A line plotted through the data points would give a very steep utility graph for $\bar{M}$. This is undesirable since a small change in a predicted maintenance time will give substantially different utility values. Also, the data does not suggest any particular form to the utility graph, even when plotted on a larger scale. It would appear sensible to construct a utility graph using a line which bounds the data as shown in Fig. 8. One stray point is omitted. Since the mean corrective maintenance times are small when compared with the calculated mean time to repair, there is little harm in allowing the utility graph for corrective maintenance time to 'spread out' in this way to make utility less sensitive to changes in repair time prediction.

(4) *Machine output*

Four of the five sets of data used fell within what was considered to be a feasible range of 0–1000 batteries per shift. They are plotted in pairs. One pair gives a linear utility graph (Fig. 9(a)), whilst the other has rapidly changing utility values over the range 600–800 batteries per shift (Fig. 9(b)). In the latter case utility increases from approximately 3.5 to 9.5 over this range, and it could be argued that this is virtually a step change from zero to maximum utility at about 700 batteries per shift. In view of the problems associated with sensitive utility graphs, and from careful consideration of the meaning of the two graphs, it was decided to adopt the linear utility graph for machine output (Fig. 9(a)). A reduction of utility value by a factor of almost three for a fall in output from 800 to 600 batteries per shift (Fig. 9(b)) is a harsh judgement.

2.3 Scale factors

Scaling factors were obtained using judgements primarily based on experience with similar equipment. Also, the relationship between some variables and the implications of the utility graphs were considered. Problems with previous 'wrap–stack' machines had been due to poor reliability. Feedback from plant had tended to be critical of

Appendix 3

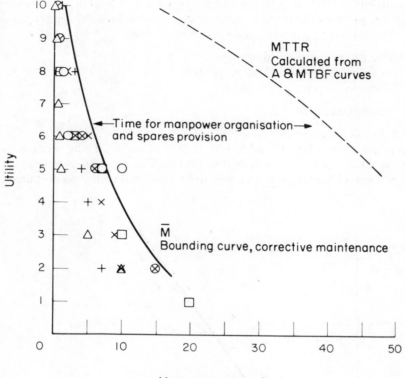

Fig. 8. **Utility graph: repair time**

reliability rather than output. It would seem reasonable, therefore, to scale reliability above machine output. The latest machine would be delivered to users who have been influenced by previous poor reliability experience, and so a good performance with respect to MTBF must be rated highly.

There is a case for reducing the weighting given to $\bar{M}$. The utility graph for this variable is very steep and the calculated mean time to repair graph has repair times some five times greater for the same utility. This gives a good deal of time for maintenance management to organize a repair compared with the time to correct the fault for a given

52 *Appendix 3*

availability. Although the utility graph drawn is lenient, as discussed above, the mean corrective maintenance time values do appear harsh.

The following scale factors were used:

Mean time between failures: 3
Mean corrective maintenance times: 1
Machine output: 2

3 Machine analysis

3.1 Detail review

In order to calculate $\bar{M}$ and MTBF for the machine, it was necessary to undertake a detailed study of the design. The machine was split into functional elements, e.g., primary drive unit, and with the assistance of

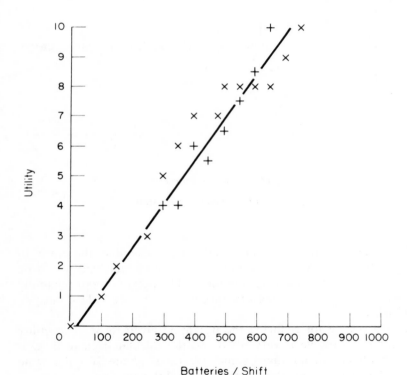

Batteries / Shift

Fig. 9. (a,b) Utility graphs: machine output

Appendix 3

a high quality fitter a component study was carried out. For each component, maintenance operations were identified and estimates of the corrective maintenance times were made. In each case it was assumed that any parts needed were immediately available, and that the repair was undertaken with prior knowledge and working in the workshop environment. These conditions were stipulated to allow concentration on the requirements for each repair without having to consider additional factors, or to 'put himself in anothers position'.

Allowances were made for plant conditions in the final assessment stage. Reliability data for each component was obtained from Green and Bourne **(21)** as a basic failure rate.

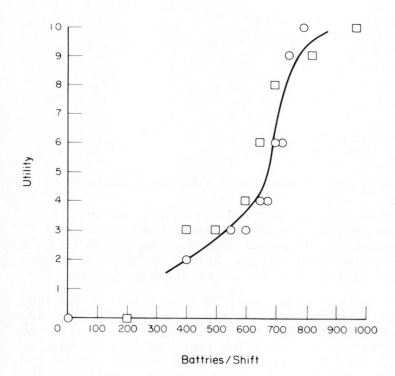

Batteries/Shift

The complete compiled list of parts with their respective repair times and failure rates is given in Table 1. The component classes considered were: elements of power transmission systems, certain adjustment screws, other moving parts, and pneumatic components. Structural failures were omitted since their probability of occurrence is small compared to the component classes given above. Where failure rate data were not available for a particular component, or where the details of a component were not completely known, a failure rate for similar components was used. In total 163 components were incorporated into the study.

3.2 Mean time between failures (MTBF)

There is no redundancy built into the machine, therefore all components can be considered to be in series

$$\text{Survival probability,} \quad R(t) = e^{(-\Sigma\lambda_i)t}$$

$$\text{MTBF} = \int_o^\infty R(t) \, dt$$

$$= \frac{1}{\Sigma\lambda_i}$$

Now

$$\Sigma\lambda_i = Z_1.Z_2.Z_3.\Sigma\lambda_{basic}$$
$$\Sigma\lambda_{basic} = 2.14 \times 10^{-3} \text{ hr}^{-1}$$

The service factors, Z, were obtained from Greene and Bourne. They were:

Environment (Z_1): 2; plant conditions are dusty and dirty
Temperature (Z_3): 1; no excesses in temperature are present
Component duty with respect to nominal rating (Z_2): for ratings of 100, 120, 140, and 160 per cent, Z_2 takes the values 1, 2, 4, and 8 respectively. The last number was obtained by interpolation.

The 'wrap–stack' machine was fitted with a variable speed control. Therefore, depending upon how fast it is run (component duty with respect to nominal rating), the mean time between failures for the machine can be calculated. This was carried out for two assumptions (A,B) for nominal output rating: 400 batteries per shift and 500 batteries per shift. The results are shown in Table 2.

Appendix 3

Table 2. MTBF as a function of machine output (batteries/shift)

Batteries per shift		MTBF
A	B	(hr)
640	800	29
560	700	58
480	600	117
400*	500*	233

*Nominal rating assumptions

3.3 Mean corrective maintenance time

This was calculated to be 1.32 hours for a design life of 25 000 hours with $Z_2 = 1$, and 1.29 hours with $Z_2 = 4$. No secondary failure modes were identified. The calculation is insensitive to the value of service factors which are, of course, constant for all components. Also, although mean corrective maintenance time is a function of design life, variations in design life (constant for all components) do not change the calculations significantly. The variation in $\bar{M}$, bearing in mind the accuracy of calculation, does not justify considering it as a function of output.

The above calculations are based on estimates for a fitter working under good conditions. Under the envisaged plant conditions of a dusty atmosphere and generally a dirty environment, these maintenance times will be exceeded. It was estimated that repairs could take up to four times as long, which gives a predicted mean corrective maintenance time of approximately 5.3 hours.

4 System utility

The above values of mean time between failures, machine output, and mean corrective maintenance time were converted to utility scores. The machine's system utility was then calculated over the range of machine outputs and for each assumption. The results are shown in Table 3.

From these results, it is possible to decide the optimum running speed of the 'wrap–stack' machine. Consider an output setting in the range 480–500 batteries per shift. If the pessimistic assumption (A) is correct, then the system utility is about 7.8. If the optimistic assumption (B) is correct then the system utility is approximately 7.9. For a higher speed of say 600 batteries per shift, the optimistic assumption gives a system utility of 8.5, but the pessimistic assumption has a system utility of about 6 (by inspection). This last figure is low, and in view of past experience

Table 3. System utility as a function of machine output (batteries/shift)

Machine output		System utility	
A	B	A	B
640	800	4.1	4.2
560	700	6.3	6.7
480	600	7.8	8.5
400*	500*	7.2	7.9

*Nominal rating assumption

with 'wrap–stack' machines, it was considered appropriate to opt for a speed range of 480–500 batteries per shift. Thus the minimum system utility is 7.8 even if the pessimistic assumption holds in practice.

5 Summary

The optimum output setting of the machine is 480–500 batteries per shift. At this production rate a system utility of 7.8 is predicted, which was considered to be a satisfactory value. A fixed output speed would be supplied to the customer to prevent the user increasing output to obtain short term benefits before breakdowns occurred.

The predicted mean corrective maintenance time was 5.3 hours under plant conditions giving a utility score of 6.3 from Fig. 8. This value is rather low and draws attention to the difference between 'what the designers wanted' and 'what was actually produced'. Several points were pointed out to the company where maintainability improvements could be made.

Typically, poor maintainability resulted from lack of attention to detail and from accessibility problems caused by putting together sub-assemblies which were designed independently.

For the optimum machine output given above, the predicted MTBF is 117 hours and the mean corrective maintenance time is 5.3 hours. In order to achieve an availability of 80 per cent, maintenance management have on average up to 24 hours to organize maintenance operations, including the provision of spares. At first sight this appears to be a long time, but when one considers that some parts may have to be imported, others made specially, this time takes on a different perspective. It should be remembered that potential applications for this equipment lie in several countries. A low number of incidences of long

Appendix 3

waits for spares will raise the average time for organizing maintenance operations. Whilst such factors as parts holding policy is outside the control of the designer, the above calculation is a useful guide to maintenance management and is a direct result of the functional unit evaluation exercise.

Appendix 4

ASPECTS OF THE COMPONENT ANALYSIS METHOD APPLIED TO PROCESS VALVES

1 Introduction

Two examples are given below: the first case refers to the 'moving parts' of a ball plug valve, and the second is a comparative reliability analysis of two types of valve. The analysis of the ball plug valve arose from a proposal to use this type of valve on part of a plant which the author examined.

2 Ball plug valve analysis

2.1 Project description

A ball mill was used to prepare powder in a particular process line. A cylindrical drum containing a mixture of steel balls and powder is rotated slowly and the steel balls progressively grind down the powder to the correct grain size.

The powder is hazardous and particularly difficult to handle. Pipes of less than 150 mm diameter are used for safety reasons, but the powder is very 'sticky' by nature and tends to block small diameter pipes. Some difficulty had been experienced in this respect when emptying the ball mill. The present emptying device, a removable plug, was also heavy and difficult to handle. It was proposed to replace that device with a 100 mm ball plug valve of plastic construction. This was considered to have a good sized bore to allow the powder to flow, to have a simple 90 degree turn between the open and closed positions (no lifting), and the use of plastic would minimize the out of balance force on the ball mill bearings.

2.2 Valve analysis

Applying the component analysis method, the first point to consider is 'moving parts'. The mating parts that have relative motion during valve operation are:

(1) ball–body;
(2) ball–ball seats;
(3) stem–stem packing.

Appendix 4

If the valve should fail for any reason, it would be removed and replaced as a whole.

Valve movements during operation were first considered. When the valve is neither fully open nor fully closed, powder will enter the space between the ball and the body. Although the life of the ball seats and to a lesser extent that of the stem packing will be reduced by the action of the powder, more important is the sensitivity of the ball to seizure within the body. In time the ball-body clearance will fill with powder and the ball will only rotate by shearing the powder surrounding its circumference.

Referring to Fig. 10, the ball rotates about X–X to operate the valve. Powder is assumed to be packed around the ball circumference and through the bore. At the port openings it is assumed that there are dead powder zones which rotate with the ball as indicated in the figure. When the ball rotates, shearing of the powder takes place around the circumference of a complete sphere.

If the sphere has a constant shear stress, τ, acting on its surface due to the packed powder, then the torque required to turn the element is

$$dT = \tau r^2 \, R d\phi d\theta$$

and for the whole sphere

$$T = 2\pi \, \tau r^3 \int_{-\pi/2}^{\pi/2} \cos^2 \theta \, d\theta$$

Without further evaluation, this simple calculation draws attention to the importance of the ball size, since the torque required to turn the valve is proportional to the cube of the ball radius. Bearing in mind the nature of the powder being handled, this gives cause for concern about the strength of the plastic valve. The valve size cannot be reduced because the passage through it would be too small. Before proceeding, and on the basis of the above calculation, non-active trials were undertaken using calcium fluoride powder which behaves similarly to the process material. Repeated opening and closing operations of the valve became increasingly difficult as powder gathered round the ball until eventually the valve stem fractured.

The plastic valve was replaced by one of stainless steel which was strong enough to shear the powder and also had the required corrosion resistance. The ball mill speed was slow and the calculated radial load on the mill bearings was within their capacity.

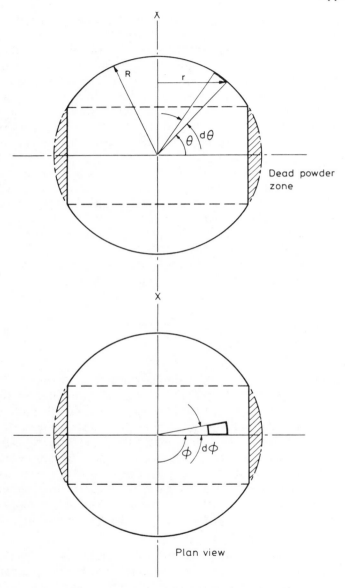

Fig. 10. Element defined on a rotating ball

Appendix 4

Although the initial cost of equipment was increased over fivefold, a much more costly exercise of valve maintenance due to failure was avoided by undertaking the detailed component study.

2 Comparative reliability analysis

The analysis compares directly two valves, A and B, which are described below. There was insufficient relevant data available for a quantitative reliability assessment, but, nevertheless, a judgement was required.

Valve A   A bonnet design valve using a replaceable seat, and with a stem sealed by a metal bellows and packing.

Valve B   A bonnet design valve using an integral seat, and with the bellows seal replaced by stem packing (giving a double stem seal).

Leak paths for valve A:

(1) bonnet/body gasket failure;
(2) valve seat/body gasket failure;
(3) disc/seat leakage;
(4) (5) bellows failure and stem packing leakage.

Denoting the survival probabilities for each element respectively as P_1, P_2, P_3, P_4 and P_5.

Elements 4 and 5 form a parallel pair, which is in series with the remaining elements. Assume active redundancy since the packing seal is always being worn by the stem even though it is not under pressure. The survival probability of valve A is then:

$$P_A = P_1 P_2 P_3 (P_5 + P_4 - P_4 \cdot P_5)$$

Valve B has an integral seat, hence $P_2 = 1$. Also, the bellows is replaced by stem packing, therefore P_4 is replaced by P_5. The survival probability for this alternative valve is then:

$$P_B = P_1 P_3 (2P_5 - P_5^2)$$

In order to compare the two designs a judgement concerning the relative merits of bellows sealing and stem packing must be made. Even without a specific duty to refer to in this case, a realistic assessment would suggest that the bellows is much less likely to leak, say $P_4 = 2P_5$.

Therefore,

$$P_A = P_1 P_2 P_3 (3P_5 - 2P_5^2)$$

and

$$P_B = P_1 P_3 (2P_5 - P_5^2)$$

Dividing the two equations

$$P_{rel} = \frac{P_A}{P_B} = \frac{P_2(3 - 2P_5)}{(2 - P_5)}$$

Now, the maximum possible value for P_5 is 0.5, since we have decided that $P_4 = 2P_5$ and P_4 cannot exceed unity. The integrity of a bellows seal is excellent, so that whilst P_4 cannot strictly be unity, a good approximation for P_5 is 0.5.

Substitution gives

$$\frac{P_{rel}}{P_2} = \frac{4}{3}$$

This is a sensible result since it predicts that, if valve B had only the bellows replaced by stem packing, but was otherwise identical to valve A, then the survival probability is reduced by 25 per cent with the above assumptions.

However, the alternative design has an integral seat, and

$$P_{rel} = \frac{4}{3} P_2 \quad \text{or} \quad P_A = (\frac{4}{3} P_2) P_B$$

For valve A to be more reliable than valve B, the survival probability of the seat gasket (P_2) must be greater than 0.75, based on the following assumptions

$$P_{bellows} \simeq 1$$

$$P_{stem\ packing} \simeq 0.5$$

If, on reflection, the result appears inconsistent with these assumptions, then the analysis can be quickly repeated.

References

(1) Ministry of Technology, *Report by the Working Party on Maintenance Engineering*, 1970 (HMSO, London. ISBN 0 11 470107 5).
(2) Comparative maintenance costs, 1980 (Centre for Interfirm Comparison, London).
(3) SMITH, D. J., *Reliability and maintainability in perspective*, 1981 (Macmillan, London).
(4) JEFFERIES, B., SADLER, J. and WILLIAMS, P. R., 'The design audit', *Terotechnica*, 1980, vol. **1**, pp. 237–241.
(5) JEFFERIES, B. and WESTGARTH, D., 'The design audit', *Chart. mech. Engr*, 1981, vol. **28**, No. 3, pp. 24–26.
(6) BAKER, G. J., 'Quality assurance in engineering design', *Engng Designer*, 1981, vol. **3**, pp. 16–18.
(7) NAPIER, M. A., 'Design review as a means of achieving quality in design', *Engineering*, 1979, vol. **219**, No. 3, pp. 286–289.
(8) LAWLOR, A. J., 'Quality assurance in design and development', *Quality Assurance*, 1978, vol. **4**, No. 3, pp. 87–91.
(9) *Design Reviews aid Truck Manufacturers and Users*. Terotechnology Case History No. 12, 1977 (Committee for Terotechnology, Department of Industry).
(10) *Engineering Design Handbook*, 1967, US Army Material Command, AD823539, (Microinfo, London).
(11) LEVESQUE, C. R., 'Design audit concept in new product development', ASME Paper No. 78-DE-W-2, 1978, pp. 1–4.
(12) JONES, J. C., *Design Methods*, 1970 (Wiley, London).
(13) JACOBS, R. M. and MIHALASKY, J., 'Minimising hazards in product design through "Design Review"'. ASME, 1973, pp. 45–55.
(14) PUGH, S., 'The design audit—how to use it'. Design Engineers Conference. Birmingham, UK, Oct. 22–26, 1979, 4A3–4A3.6.
(15) de NEUFVILLE, R. and STAFFORD, J. H., *Systems analysis for engineers and managers*, 1971 (McGraw-Hill, New York).
(16) CHATFIELD, C., *Statistics for technology*, 1970 (Chapman and Hall, London).
(17) SMITH, C. O., *Introduction to reliability in design*, 1976 (McGraw-Hill, New York).
(18) SIDDALL, J. N., *Analytical decision making in engineering design*, 1982 (Prentice-Hall, New Jersey).
(19) THOMPSON, G., 'The evaluation of a functional unit in a design audit—system utility', 2nd Cairo University MDP Conference, Egypt, Dec. 1982, pp. 337–344.
(20) GREENE, A. E., 'Reliability prediction', *Proc. Instn. mech. Engrs*, 1969–70, vol. **184**, pp. 17–24.
(21) GREENE, A. E. and BOURNE, A. J., *Reliability technology*, 1972 (Wiley, London).
(22) QUIRK, G. G. 'Logic Design Factors. An Approach to predicting Mechanical Design Reliability', Report No. R61 POD6, 1961, Ordnance Department, Defence Electronic Division, General Electric Co., Massachusetts.